FLEETING MOMENTS *of* FIERCE CLARITY

Journal of a New England Poet

FLEETING MOMENTS *of* FIERCE CLARITY

Journal of a New England Poet

L.M. Browning

HOMEBOUND PUBLICATIONS
Independent Publisher of Contemplative Literature

PUBLISHED BY HOMEBOUND PUBLICATIONS

Copyright © 2012 by L.M. Browning
All Rights Reserved

All Rights Reserved. Without limiting the rights under copyright reserved above, no part of this publication may be reproduced, stored in or introduced into a retrieval system or transmitted in any means (electronic, mechanical, photocopying, recording or otherwise) without the prior written permission of both the copyright owner and publisher except for brief quotations embodied in critical articles and reviews.

For bulk ordering information or permissions write:
Homebound Publications, PO Box 1442
Pawcatuck, Connecticut 06379 United States of America

John Haines, excerpt from "Of Traps and Snares" from *The Stars, The Snow, The Fire: Twenty Five Years in the Alaska Wilderness*. Copyright © 1989 by John Haines. Reprinted with the permission of The Permissions Company, Inc., on behalf of Graywolf Press, Minneapolis, Minnesota, www.graywolfpress.com

Contains excerpts from *The Nameless Man* by L.M. Browning and Marianne Browning. Copyright © 2011 by L.M. Browning

The poems: *Departure from Avery Point, Summer's Passing* and *Autumnal Evening* previously appeared in *Written River: A Journal of Eco-Poetics*

FIRST EDITION
ISBN: 978-1-938846-01-4 (pbk)

BOOK DESIGN
Front Cover Image: Rain Over Street Lights © Silviu Marcu
Title Page Image Attribution: © Shuttershock.com / CarpathianPrince
Cover and Interior Design: Leslie M. Browning

Library of Congress Cataloging-in-Publication Data

Browning, L. M.
 Fleeting moments of fierce clarity : journal of a New England poet / L.M. Browning. -- 1st ed.
   p. cm.
 ISBN 978-1-938846-01-4 (pbk.)
 I. Title.
 PS3602.R738F57 2012
 811'.6--dc23
                                                            2012036056

10 9 8 7 6 5 4 3 2 1

ALSO BY L.M. BROWNING

POETRY

Ruminations at Twilight: Poetry Exploring the Sacred
Oak Wise: Poetry Exploring an Ecological Faith
The Barren Plain: Poetry Exploring the Reality of the Modern Wasteland

FICTION

The Nameless Man

CONTENTS

Foreword by Ian Marshall	ix
Introduction	xv
Pieces	1
Awakening	5
Revolutions	9
The Acts of the Powerless Ones	13
Roots in the Sea	17
The Cure for Blindness	19
Autumnal Evening	21
A Progression	23
Summer's Passing	27
Across the Distance	31
Luminaria	35
The Late Train Home	39
During the Long Day, Over the Sacred Night	41
Departure from Avery Point	45
The Lament of the Wayfarer	51
The Lowest Point	53
The Truce	55
Withered	59
Blue Mornings on the Concord River	63
On the Far Side of Walden	67
Self-sufficient	71
The Boat Rocker	73
Fleeting Moments of Fierce Clarity	75
Deep Notes, Quick Pace	79
All or None	81
The Voice	85
Authenticity	91
The Mystic River	93
Simplicity	95
Conclusion	97
Closing Note	101
Acknowledgements	102

FOREWORD
by Ian Marshall

Where the Journal Meets the Journey

L.M. Browning introduces her project in *Fleeting Moments of Fierce Clarity* as a poetic travel journal set in her home geography of New England, which might make one curious about the link between the words journeying and journaling. Clearly there is some etymological common ground that constitutes the terrain that Browning is exploring in her spiritual travels. A basic recollection of high school French might bring back to us the fact that the word for "day" is la jour. Coming to our language via Old French, the word journey originally referred to a day's travel, and a journal (related to the Latin word diurnal, meaning daily) was a written record of daily events, perhaps deriving from a journey-book, or the log of a trip. Of course, these days we no longer think of either a journey or a journal being limited to the events of a single day, and that shift came when people would speak of an extended trip as "several days' journey." How interesting that these words that we think of as opposites—one taking us away from home, the other an intimate recording of our innermost thoughts—have the same starting point.

What Browning does in *Fleeting Moments of Fierce Clarity* is reconcile the apparent opposites implied in journal and journey, offering a series of waypoints in an intensely personal spiritual quest where the path leads, paradoxically, both outward and inward. I'm reminded of John Muir's famous line, "I only went out for a walk, and finally concluded to stay out till sundown, for going out, I found, was really going in." But perhaps the model for Browning's journey can be found a little closer to home, in Henry Thoreau, who not only provides one of Browning's epigraphs but whose spirit infuses many of these pages. When Browning says "You do not need to go to the edges of the earth to learn who you are, only the edges of yourself," we might think of Thoreau's dictum that "It is not worth the while to go round the world to count the cats in Zanzibar," and that we should "obey the precepts of the old philosopher, and Explore thyself." It may be that, as Browning says, "we must work a little harder to feel that sense of wonder" when we travel close to home, but "rediscovering the beauty of what has become ordinary has its own sweetness" when we perceive "the dearness in what might otherwise be regarded as mundane."

Like Thoreau, Browning in these poems is tackling, among other big questions about existence, the Thoreauvian problem of how to live. For her the deepest satisfactions come from somewhere other than the office or factory: "Working less," she says, "gives us more meaningful time out of the workplace to cultivate a richer inner-life." She argues for the value of simplicity, offering in her poem "Self-sufficient" an accounting of the things we wait and yearn for that are outside the self, the things we mistakenly think will provide meaning and clarity: God, lover, father, mother, savior, answer. This is a very spiritual book, full of the author's "longing to understand the greater matters." Her great theme is spiritual renewal, and poems like "Authenticity" work through the process of finding, asserting, and being oneself—in Browning's case, an "old soul that lives in [a] youthful body." The poems reveal Browning's interest in religious questing, and the language and rhythms seem to owe something to various scriptures. "Revolutions" could be from the Old Testament: "Those who proclaimed themselves king / shall, in the turning, / wake to find themselves / the servant to their servant." Poems like "During the Long Day, Over the Sacred Night" offer paradoxical truths that sound like something out of the Tao: "There, / in the dark, / when your light / was no longer upon me, / I could see myself." In "A Progression" the litany of binary oppositions sounds like either Biblical repetition or Tao-like paradox, but in either case the rhythmic influence seems like something scriptural: "In one life I was a warrior / and in another I was a pacifist"; the list of past lives goes on to include doctor-patient, reader-writer, philosopher-fool, virgin-lover, servant-master, idealist-cynic, proud-self-loathing, and willful-swayable. Perhaps here too we also see influences of Thoreau again ("I left the woods for as good a reason as I went there. Perhaps it seemed to me that I had many more lives to live...") or of another of Browning's epigraph-sources, Walt Whitman: "Do I contradict myself? / Very well then I contradict myself. / (I am large, I contain multitudes.)" Browning too is penning a song of the spiritual self.

The spiritual themes and influences of Browning's poems reflect a didactic impulse: she wants us to think about the way we live—and given the moral insufficiencies of too many of our lives, to change. She is offering instructions on how to be. Often those instructions come in neat aphorisms: "the body need not die for one life to end and another to begin"; "To gain knowledge of the Divine / one must leave behind all religion"; "true civilization / is found in the places man / has yet to touch"; "A stable house cannot be built upon the shifting sand." Thoreau too was a master of epigram, and many of these apothegms are in the vein of Thoreau's rhetorical trick of inverting common wisdom in search of a deeper spiritual truth. There is also a strong element of parable and allegory in Browning's poems, as in "The Lament of the Wayfarer": "When the day comes / and I at last clear this dense wood, / I shall meet you on the other side. . . . When the day comes / and I can at last pull up / the moorings holding my soul in place, / I shall journey to you / and set fire to this vessel I dwell in / Never to leave you again." These deep woods and unmoored vessels, we are aware, are always something more than literal. There is also a quick parable in "On the Far Side of Walden" when Browning says that "Ankle-deep in mud / along the banks of Walden, / I find my footing"—an innocent seeming description that becomes a mini-allegory about returning to the earth to get our inner-selves in touch with the real. The practice seems reminiscent of Thoreau's transcendental method, whereby, as Emerson put it, "natural facts are symbols of spiritual facts." In the same poem Browning writes, "Coming to the end of the path / I am not who I was / at the beginning," which makes that path take on metaphoric resonance as something more than a literal footpath around a pond. In "The Boat Rocker" the allegory relies on the way Browning takes the familiar phrase and makes it literal. These stories in free verse remind me of what Mark Turner says in *The Literary Mind*: that parable—the ability to see an event or even an observation of a scene as meaningful and to project its meaning on to your own experience—is the essence of literature and of human cognition.

Often the lessons Browning conveys have to do with the nature of time. In "Autumnal Evening" she describes the conversion of a living tree into log, flame, and ash: "The red kernels of the burning oak log smoked, / blessing those who stood around, / witnessing the cremation of its century-old life." These lines give a sense of time's magic and the awe we feel at how it can accumulate and be lost. At times the awe is reserved for moments of sudden connection to the long ago, as in "Blue Mornings on the Concord River": "As I grasped the old wrought doorknobs, I shook hands with the past." And in "Deep Notes, Quick Pace" the awe is aimed at the fleetingness of time, what the haiku poet Bashō called fueki ryūkō, the unchanging and the ever-changing: Browning speaks of "willing the moment to last, knowing that it will fade." There is a sad poignancy there, reflecting on the impermanency of all things.

Browning is at her best as a poet when the parabolic dimension is most firmly grounded in image, with more appeal to the senses than mind or soul, as in "Blue Mornings on the Concord River." The poem offers a haiku-like juxtaposition of images, the human and the natural brought into conjunction with one another, and it's another poem where the haunt of the past becomes suddenly and dramatically present . We have the past rendered in this scene: "Running through the draping grasses / the farm boys fired their muskets. / Son against son, / as blue collided with red, / the world changed." That world and event still seem present in this description near the end of the poem: "The field is smoking / as the morning fog rises." In poems like this Browning is simply describing scenes experienced, seeking "moments of clarity" that, as she says in her introduction, may "yield no philosophical insight" (at least not one that is explicitly stated) but in which "everything sharpens."

My favorite poem here is "Departure from Avery Point," another one packed with images, where the clarity of pure seeing is the meaning and the lesson of the poem. All Browning takes in via the senses builds up til "The isle I am on / merges with the isle on the chart." When the "duties on the mainland call," we can feel her reluctance in the unfinished thought that closes the poem: "And I, unable to go deaf." We understand that she'll have to heed that call and leave the island, leave the state of pure being, the fleeting moment of fierce clarity. Because we can never stay there, can we? We must always be journeying (and journaling) on.

INTRODUCTION
A Note From The Author

The Idea of a Travel Journal

This collection was gathered out of a desire to create a poetic journal chronicling my various journeys throughout New England. When assembling this book, there were times when I asked myself if my putting together the equivalent of a travel journal wasn't a self-indulgent act. I mean, I am hardly a globetrotter. Born to limited means, my passport is as blank as it was the day it arrived in the mail. So, given this, what gives me the authority to compose a book boasting to be a *travel journal?*

As readers, we live vicariously through the adventurers of our generation. We read the chronicles of those who left the comforts of home to strike out into the untamed and unknown, and through absorbing their experiences we are emboldened to heed our own yearnings for new landscapes. Society seems to have subconsciously adopted this notion that in leaving behind all that we have ever known, we will find ourselves—that there, at the ends of the earth, each of us can define the edges of ourself. I think this is an unrealistic ideal.

Our imagination is sparked by those travelers who set off with reckless abandon. Yet for so many of us there is a reality gap between the life of those we follow on the page and the life we ourselves must lead. The 9-5 job hardly supports our basic survival let alone the heights of our dreams. We work from the time we rise to the time we go to sleep just to support the basic needs of our body, all the while having to neglect the needs of our soul.

People speak of long pilgrimages as a rite of passage. The path through the Holy Land, the Way of St. James, the pilgrimage to Mecca, the Appalachian Trail and so on. I have never followed a map from one side of a country to another, but I have made my journey. Four pairs of leather boots worn through and 10,000 miles later, I have endured the long path.

For the majority of my life I have been hard-pressed to keep food on the table, leaving the possibility of traveling abroad ever a dream. Not all of us are able to set foot upon the far-off lands that call to us. While the number of destinations I dream of one day going to number into the dozens, my bank statement does not support the breadth of my aspirations. Do not think I am using lack of money as an excuse to stay in my comfort zone; I am not. Rather I am facing a hard truth of circumstance: Not all of us have the means to pick up and travel to different countries while heeding that desire to *find ourselves*. In these hard financial times, the majority of us must *find ourselves* while sticking relatively close to home. Leading me to ask: Must we go to the ends of the earth to gather the strands of our identity?

Every Journey is a Pilgrimage

The purpose of a pilgrimage is about setting aside a long period of time in which the only focus is to be the matters of the soul. Many believe a pilgrimage is about going away but it isn't; it is about coming home. Those who choose to go on pilgrimage have already ventured away from themselves; they go on pilgrimage as a means to journey back to who they are.

Many a time we believe we must go away from all that is familiar if we are to focus on our inner-wellbeing because we feel it is the only way to escape all that drains and distracts us so that we can turn inward and tend to what ails us.

For personal reasons, I could not go to foreign lands when I felt the need to make a pilgrimage unto myself. So instead, I walked the same roads I had since I was a child and arranged my life itself as a period of time in which the only focus is to be the matters of the soul. All that was detrimental that could be left behind, was. I broke ties with everything and everyone that insulted or confined my soul, allowing me to go forward and find my path into a healthier way of being.

If I could pass along one wish to you—heed you to do one thing—it would be: Make your life the pilgrimage—make your life the time of contemplation, of growth, and of returning to that place of authenticity and innocence, wherever it may be.

Unable to go outward, I went *inward.* The radius of my physical world so limited by circumstance, I spent many years walking the internal landscapes. When at last I was able to "loosen the belt" a bit and stretch the legs of my stiffened dreams, I found myself exploring, not foreign countries, but the rich country of New England, of which I am a native daughter.

No matter where I am situated on this earth I think I will always be a bit of a homebody, and happily so. This is not to say I spend my days cooped up away from the sunlight; rather, that I appreciate my home as a sanctuary that I am able to create and enjoy. I find peace in simple things. Having endured periods of homelessness during my childhood, I have come to appreciate my small apartment along the Connecticut coastline more than anything.

Of course, in spite of my contentment at home, I do indeed have times of restlessness. The wanderlust strikes and I feel the need to enter an inviting new surround. Working within my means, I cannot pick up and backpack through Europe when these feelings strike. For several years I felt denied life-defining experiences by my meager income. But like so many things in this life, it is all a matter of *perspective.* There is a difference between not being *able* to go on a fantastical, far-off trip to find one's self and not *needing* to do so.

We do not need to go to the edges of the earth to learn who we are, only the edges of ourself. Nature aids us in turning within yet it need not be a foreign landscape.

Travel freshens the senses. A feng shui of the horizon, when we leave behind the familiar our renewed curiosity widens our eyes and we take in all the little details of our new environment. We each seek change but there are times when our life does not allow us to see to our inner-wellbeing.

In these times, when I cannot simply pick up and go, I make do with a walk about my hometown. When in the confines of our local community, we must work a little harder to feel a sense of wonder; for sadly, when we see a thing daily, its beauty fades into the background and become mundane. Nevertheless, rediscovering the beauty of what has become ordinary has its own sweetness. Seeing anew the beauty of what we have gazed upon each day, which has become tired to us—this is a revelation.

After all, what was Walden Pond before Thoreau chose it as the place for his introspection? When he chose to go off on his own into the wild and reflect, he did what was within his means. He lived off a small plot of land owned by Emerson, along the banks of a pond just outside Concord—his hometown.

This Composition

Normally I write spiritual verse spurred by some longing to understand the greater matters at work in my life. However, many of the poems I write while traveling describe a scene experienced, not a spiritual concept. Some moments of clarity yield no philosophical insight; rather, the fruits from such journeys are mentally medicinal. When once I was so consumed within my own thoughts, work, and worries that I could not take in my own surroundings, everything sharpened. Becoming aware of the dearness in what might otherwise be regarded as mundane is the ultimate form of insight.

Mingled within my poetry written during my various New England wanderings are those pieces penned in the solace of my home along the Connecticut coastline. These verses are sketches of the internal landscape, I walked during my long nights spent in reflection. While these entries may not seem to fit into this collection of travel writing, I believe night to be its own journey. Relieved of our daylight toils and duties, our suppressed spirit can breathe at night and roam the different landscapes of the soul, mind, and imagination.

The collection opens with a number of poems written during my times of contemplation at home. Then, slowly, as my world widened and I found myself taken down new roads, the focus of the collection widens to include my travels around New England.

Each poem in this collection features a brief journal entry penned about the same time as the verse. Some of these introductions concern spiritual realizations I experienced around the time; while others are memory-driven.

In contemporary poetry circles, my decision to include journal entries as introduction to the verse would be questioned. A poem, it is believed, is meant to *speak for itself*. Introductions are thought to narrow the reader's ability to interpret the poem. Opponents of my style would argue that, in giving you my interpretation of the poem's meaning, I have made it so your mind will not venture beyond the boundaries I have set. However, I believe that when a writer shares the epicenter of a poem—the place in their life from which it emanated—it can give the reader a *relating point* with the poet. As human beings we struggle with similar problems each day and, though we feel isolated in what we endure, we actually handle our struggles in quite the same manner—we feel the same emotions and seek the same answers to the same questions.

When a poet hosts a reading, before presenting each poem to the audience, it is customary to give a brief background on the origin of the piece. I find that, when I hear a poet speak of the emotions and experiences that are at the root of the poem, I feel intimately connected with the piece as it is being read. I can still take the words into my own context and relate them to my own life but, simultaneously, I am sharing in a private moment in the poet's life and feel bonded with them.

The poems within this collection have a different meaning for me than they will for you—the reader. The poems, for me, embody *fleeting moments of fierce clarity* as they capture those brief, transcendent experiences of pure love and appreciation that, when experienced, affirm to us that the struggles in our daily life are worth persevering.

— L.M. Browning
Connecticut, Summer 2012

"The trails I made led outward into the hills and swamps, but they led inward also. And from the study of things underfoot, and from reading and thinking, came a kind of exploration, myself and the land. …to take the trail and not look back."
— John Haines *The Stars, The Snow, The Fire*

"I have a great deal of company in my house; especially in the morning, when nobody calls."
— Henry David Thoreau

"I bequeathe myself to the dirt,
to grow from the grass I love;
If you want me again,
look for me under your boot-soles."
—Walt Whitman, *Leaves of Grass*

Pieces
Connecticut, August 2009

The precise age of one's soul is something only the individual can know. Few of us lead a single life. There are times when I find myself longing for some nameless thing—something my soul once knew that is no longer present. I mourn for loved ones lost whose names and faces I cannot recall. I feel a pull towards places upon which I have never set foot in this body but somehow still know intimately. And in these yearnings I experience whispers of my own unknown past.

There is one who knows more about my existence than I do. There is one who has watched my progression; not just the growth of this body but of this soul—a silent steady parent who is guardian over all lives, not just one.

The mind in this body is new; it holds not the memories of all that has past before its creation. Nonetheless, deep within the soul the knowledge of what has been lives and from time to time the soul whispers of what it knows.

I have learned over my path that the division between lifetimes can be drawn by many an occurrence. Sometimes by the death of the body and sometimes by an evolution of the soul.

Many think that only death can draw the line between lives but many lifetimes can be lived in a single body. The body can endure, yet lifetimes can end and begin again. Our feet can travel the same paths each day yet immeasurable distance can be crossed. The physical measures of miles and years exist, yet the internal measures are what truly define the extent of what has been.

Eventually the measurements imposed by man are too limited and only the heart can define what has been endured, what has been accomplished, the distance traveled, and how many lives have been lived.

I believe in past lives. I don't necessarily believe that each of us is automatically reincarnated back into this world after our death; rather, I believe that some of us choose to return to this place—to live another life here—for whatever reason...to serve whatever purpose.

I also believe that *death* must be redefined. Yes, there is the death of our body, but the body need not die for one life to end and another to begin.

Without the death of the body, one life can end; all ties to that life, be resolved; all questions accumulated over that life, answered and all ways of living, changed. Some will live only one life, choosing to remain stationary—the same person now that they were decades ago. Yet those who seek and learn and grow shall inevitably live several lives over the course of their existence.

We need not die in body for a defining distinction to be made. We pass over a divide—an event that draws a line. The memories of the old life fade—washed out by the sun over the long distances we travel. We come to a place wherein we look back at what we once were—at where we once were—and it seems a dream...hazy and scarcely possible.

Old acquaintances revisit us like ghosts from the grave. They come to us—drawn to a face of one they recognize—looking for reunion only to greet a stranger; for, while our image is the same, the soul in us has evolved. Reborn again and again, we shed the skin of the old life as we progress; for it is too small to contain the evolving soul.

...Perhaps one day I shall be led to a *unifying moment* where I shall see the breadth of my existence—no division between lives, only a single seamless stream of being. Where all is recalled and the whole that all *the pieces* make can be appreciated.

Pieces

The hardest story to understand
Let alone convey to another
Is our own.

I do not know
Where my story began.

I cannot now recall
My original birthplace

In that first lifetime
Now in the past—forgotten.

I cannot recall in what age
I first came into being.

Nor upon what plot
My first breath was drawn.

...Cradled in whose hands?
...Christened what name?

These chapters of my soul's existence
Have been lost with time.

I search for them—
Riffling through the papers
Of my scattered mind.

But still cannot find those leaves
That speak of my past—

The knowledge that would give
All that I do know
A greater context.

I cannot start at the beginning
And I do not know the end.

And so,
You must take me as I am
   —Take my words
   And the thoughts I voice to you,
And piece them together as a mosaic

The picture of which
Might give you
Some vague impression
Of the old soul that lives
In this youthful body.

Awakening
Connecticut, July 2010

I have spent years pondering *the whys,* the questions at the center of our existence. Hard circumstances withdrew me from social circles. Unfashionably honest rather than accommodating, there was a period in my life when I wasn't invited out much. During these years, while I kept my own company, I ventured *within.*

I like to think I consciously *chose* to make the greatest portion of my life within myself rather than was confined to do so by limited choices. There are those times when I feel like Thoreau—sensing the depth there to be explored within the quietude, and so I walk the landscape of the soul rather than that of the city. Still, at other times, I feel confined like a prisoner bound to a solitary existence with nowhere else to go but inward.

Normally after we graduate from high school the world opens to us—college, career, travel, venturing out on our own. Difficult circumstances deprived me of those opportunities. After graduation the radius of my world shrank rather than expanded. During those years I read and pondered and evaluated my life, my faith, the path humanity has taken, my identity…everything.

Almost ten years passed this way. Eventually the hardship began to ease, inner-world widened, and balance came. More seemed lost than was gained during those years of solitary thought. I was disillusioned of the religion under which I was raised as well as each subsequent faith I had attempted to adopt. I had lost or parted ways with nearly all whom I called *friends* or *family.* And while I felt the grief of this accumulated loss acutely, in the end, I console myself with the truth that all I lost was an illusion and while seemingly little had been gained, what I did have left in my life was true.

We are all driven by something; my life is given momentum by my curiosity concerning the mystery of the unseen. My life seems bound to unseen workings. I began my spiritual exploration where most do: with religion. But I have long-since learned that to discover the nature of the divine we must do so firsthand—through our own progression of faith.

Over the course of fifteen years of introspection, I have deeply considered the evolution of the individual's interaction with the unseen and I believe we each make the following progression:

- Religion
    We hold a rigid doctrine and worship the divine from afar.
- Mysticism
    We are still steeped in doctrine yet are willing to consider the concept and practice of connecting directly with the divine rather than via the interface of the church.
- Spirituality
    We achieve independent thought. We remove most interfaces. We set out to gather truth firsthand. We hold a blend of religious concepts and unique beliefs.
- Individuality
    We hold fully unique ideas. We commune and connect with the divine without any interfaces. We have begun developing a personal relationship with the divine.
- Reality
    We move beyond pondering, mythos, and theory and live our belief—live aware of our part in the unseen world.

While I do indeed bear more questions than answers, I carry on, trying to live with the hope of finding further answers. Though at times my years of learning seem past, now and again days come and I realize a new truth that I have been building towards for years without even knowing it.

Even after the mind has forgotten our questions or given up on finding the answers, our heart holds our questions close and, even without our awareness, searches for the answers.

The truths I have, though they be few, are solid and hard-earned. Likewise, I have been blessed to meet new friends and kin and, while I can count their numbers upon one hand, I know they are true.

We strip ourselves down, grieve for the loss, endure the emptiness, and are slowly rebuilt one thin layer of truth at a time.

Awakening

How does one heal
From disillusionment?
   —They shed the lies
   They have woken to
   And commit themselves
   To finding that which is genuine.

How does one heal
From betrayal of self?
   —We take up the vow
   To practice fidelity to our original self.

How does one overcome doubt?
   —They become the monk.
   Devoting themselves
   To the path of believing.

The cure for the wound
Must come from the blood
Of the wound itself.

Feeling the stiff blade of the sharp razor
Move across my scalp
   —The long locks of my
      Honey hair falling away from me—
I am relieved of the burden of my vanity.
Now able to see the beauty of my soul.

Feeling the heavy cotton robes
Wrapped around me

 —The homespun linens wound
  Layer upon layer round my body—
The cynicism that came about
From my over-exposure is quelled
And I am comforted.

With the tiger-eye mala
 —One-hundred and eight beads
  Draped round my arm—
Coiled round my wrist,
I am no longer alone;

For this is my wedding band
Showing my fidelity
To my true self.

We cannot teach others through sermon,
Only through speaking candidly.

We cannot show others the way,
We can only attempt to find our own.

We cannot bolster the faith of others,
We can only keep our own robust
And let others gather from us what they will—
 Like a tree that rains down fruits
 That can be gathered by passersby
 Or dwell in the surround.

We define ourselves through belief
But more specifically
Through right action.

Revolutions
Connecticut, July 2010

Revolutions is a poem of wide scope. It is a coming to terms for victims of atrocity. After being grievously wronged we tend to fixate on what has been done to us. It can be difficult to move beyond hard truths to restore the balance that has been lost.

As one who has been deeply wounded, I know what it is to get lost in the *hows* and *whys*. For many years, following an attack committed against me, I was trapped. I reeled in all that had been done to me. I saw those who hurt me go on their way—unaffected—and it ate away at me. I didn't want revenge, I wanted some kind of justice—a restoration of the balance they disturbed.

During this time of suffering, I became acutely aware of how greatly the victims of this world suffer. So many of those injured in body and soul are voiceless—lacking a protector willing to stand for their rights.

As sickening as it is, many of those who inflict the deepest pains in this world are protected by their office or financial standing. Their hands are bloody but they are beyond reach of our easily-swayed system. So how do the victims of these corrupt ones ever find peace, knowing there will be little to no punishment for the unspeakable crimes committed?

I found my peace in the following poem...in the natural justice that history doles out over time.

Revolutions

In the turning of the Great Wheel,
Which no human being can influence nor halt

Any evil that has,
    In the course of history,
Wiped away that which was pure
Shall themselves be wiped away.

Any evil that has undone
That which is innocent,
In the turning, so too,
Shall themselves be undone.

Any evil that has conquered,
In the turning, so too,
Shall they be conquered;

Any evil that has
Uprooted the rightful way
In the turning, so too,
Shall they themselves be uprooted.

In the turning of the Great Wheel,
Which no human being can influence nor halt,

Those who have struck others
When in a position to nurture,
      Shall, in the turning,
Have their hands tied
And find themselves the helpless ones.

Those who proclaimed themselves king
        Shall, in the turning,
Wake to find themselves
The servant to their servant.

Those who hoarded wealth
        Shall, in the turning,
Wake to find themselves
Beneath the lowliest beggar.

Those teachers
Who vainly gathered students
In a desire to be revered
        Shall, in the turning,
Wake to find themselves
The student to the fools.

In the turning of the Great Wheel,
Which no human being can influence nor halt,

Any seat of power created
        Shall, in the turning,
Be forced to submit to
The one permanence: Change.

For mankind is not among
The immortal in this world.

All that we do,
        Be it good or wicked
And all that we are,
        Be it loving or evil,
Is to only last for the briefest time,
In-between the turnings of the Great Wheel.

The Acts of the Powerless Ones

Connecticut, July 2010

I have learned during my time that those with *nothing* know the worth of *everything*. I was raised to appreciate everything—to take nothing for granted.

Concepts of value seem backward in this age. As a society we are caught in a vicious cycle of selling what is priceless for what is worthless. The more materials we consume, the greater the debt we owe to the seller. If we learn how to be self-sustaining, we will consume less as a civilization and have more. Through redefining necessity we can bring ourselves back to basics and reclaim some of our precious life now given to our employer.

A smaller material life leads to a richer internal life. We are forced to trade hours of our life whenever we need something in this money-driven world. To wear the latest fashionable coat for $200.00 we give 25 hours of our life to a low-paying job in order to earn enough to buy it.

We have this life—this precious, singular, miraculous, fleeting gift—and we have given it an insulting price tag. Every day, when we go to work and clock in, we are selling our life for $8.00 an hour. The sheer staggering, horrific waste of such a system is mind-boggling.

Each day we go to work, eyeing the clock—willing the time to go faster so that the day will be over and we can go home. But what we seldom are conscious of is, as we sit there willing the clock to move faster, we are willing our life to pass us by. We are willing the hours to go quickly—hours that we will never get back—never experience again.

So many great teachers, holy men, and gurus have told us to simplify—to walk away from material possessions. One might ask:

But why? Why do I have to give up all my possessions in order to come to understand the higher power at work in this world?

We work at soul-wrenching, draining, pointless jobs for hours on end. We build nothing; we have no emotional investment in our work and wonder why we are left empty. We buy more material possessions in an effort to fill that emptiness and in so doing must continue to work the menial position that leaves us unfulfilled. This is the vicious cycle in which many of us are currently stuck. The only way to break it is to ignore the mainstream media, which runs commercials 24/7 projecting things on to us that we don't need, redefine our true necessities and break the consumerist lifestyle. We must become self-sustaining. What does this mean? It means needing fewer materialistic items to get by daily. It means bringing what we need to survive out of ourselves each day rather than going outside ourselves to attempt to buy happiness.

To be free of this detrimental cycle we must: Reevaluate our needs. We must redefine our necessities. Through doing this we will lower expenses and reclaim our life. Lower monthly expenses means we need not work as many hours to reach our required monthly budget. Working less gives us more meaningful time out of the workplace to cultivate a richer inner-life, which in turn means that mentally, physically, emotionally, and spiritually, we are healthier.

The Acts of the Powerless Ones

The refugees,
Who have carried
Their simple dreams of freedom
Over the long path that broke open their feet
And flagellated their pride,

Shall be those who build a home
For the forlorn souls in this world.

The victims,
Who have been the object
Of ill-doers, who have spent years
Untangling their mind and tracing back
The pure root of their manipulated emotions,

Shall be those who understand
The psychology of the human being.

The restless seekers who,
After passing through each religion,
Found no definitive truth,
And in the end passed into their own self

Shall be those who reveal unto us the features
Of the ever-elusive face of God.

The disavowed ones
Who left behind all known acquaintances
And the comfort of the familiar,

Taking only a few belongings in hand
And their deep-set dreams and morals,

Shall be those who teach us
What it is to live for a cause.

～

From the heart of the unjustly imprisoned,
    As they reconcile themselves with the acts
    Of those who stole their freedom,
We shall learn what it means to forgive.

From the will of the forcibly-displaced,
    As they recreate the world taken from them
    Within the small space they are restricted to,
We shall learn what it is to endure.

From the realizations of those whose stomachs
    Have been eroded by hunger's acidic waves,
    We who have an abundance yet remain empty
Shall learn what truly nourishes mankind.

From the longings of the orphan,
    As they ache for the familiar embrace
    They have never known in this life,
We learn what it is a mother and father truly are.

～

If you seek to know the worth of a thing
Go unto one who does not have it
And ask them to speak of what
It would mean to them to receive it.

Roots in the Sea
Barn Island, Connecticut July 2010

There is a hidden away spot in Stonington, Connecticut named Barn Island. It is a tidal salt marsh along the shores of Little Narragansett Bay—an inlet of the Atlantic Ocean converging with the estuary of the Pawcatuck River along the southeastern border of Connecticut and Rhode Island.

Barn Island is, in my opinion, an undervalued treasure of New London County. The preserve is a network of winding blue waterways cutting through thickly-reeded banks, leading visitors along untouched shores.

Mainly frequented by fisherman and local naturalists, the paths are rather overgrown in places; nevertheless, on a cool autumn day there is nothing better than a hike along the shore and a chance to take in some sea air.

We [myself and Andy, my dearest friend and unnamed companion throughout many of my New England rambles] had gone for a hike along the paths one afternoon, unwisely we choose a rather hot day in mid-July to make this venture. On that day, the hike was short and the picnic long. We managed to enjoy the walk despite the inferno. Andy cheerfully put up with my shutterbug habits and naturalist observations as we meandered along the coastal trails.

The embanked trails cut through the salt marsh acting like bridges between bits of wooded land. It was as we passed through one of these wooded points that I began pondering the life of the trees around us and how, unlike their land-locked fellows just inland, the roots of these trees drank in the Atlantic. Coursing through the fibers of these trees—circulating through their rings—are the tides of New England.

Roots in the Sea

The channels of sea
Meander through the dense reeds.

Winding embankments
Scallop the coast.

The paths through
The salt marsh
Leading out to open waters.

The winds linger here
In the wooded coves
Before crossing the Atlantic

Sauntering for a time
Among the oaks
Whose roots soak in the brine.

The Cure for Blindness

Night Train from Connecticut to Washington DC, Autumnal Equinox 2010

There is a bar-less prison in which the fearful dwell.

We are told that it is a good thing to leave our *comfort zone* from time to time—as if we are resting comfortably in our quiet despair and pained resignation. Traumatized, lethargic, hopeless, and afraid of change—even if it is good—we become prisoners of our claustrophobic comfort zone.

Some would say we are self-imprisoned—that we hold the key to this bar-less cell of fear and inhibition, but at the same time, it is never that simple. We are not born self-imprisoning; on the contrary, we are born open-hearted and fearless to a fault. It is the unanticipated hardship we encounter that leaves us too afraid to live our life—too traumatized to venture out of our small sphere where all is familiar and predictable, safe, and nonthreatening.

Many of us who live restricted lives do so, not because we have lost the desire to live a fuller life in a wider sphere, but rather, because we lost the confidence necessary to do so.

The Cure for Blindness

Every now and again
The eyes must see
Different sights.

Change prevents blindness
And numbness.

When we believe
That we have seen
All there is to see,

A gray curtain drops over the eyes
And the extraordinary around us
Becomes hard to make out clearly.

Yet when we leave the familiar
The lazy scenes are roused.

Inquisitively and with a sharp eye,
We approach the new part of the world
Opening to us.

Ears pricked,
Eyes clear,
Senses extended,
We wake to learn
We have been lulled into
A deep sleep of disinterest.

We must go away,
So that we can come back
And see the ordinary, anew.

Autumnal Evening

*Pennsylvania, Pocono Mountains,
Autumnal Equinox 2010*

Occasionally this New England girl ventures outside the Northeast. In 2010, I was invited south to visit a friend. This venture was a lengthy one, including stops in Washington D.C. and Virginia, finally ending with a gathering in the mountains of northeastern Pennsylvania to celebrate the autumnal equinox. During this three day stretch, my life was brought to a needed pause. The trip came at a particularly busy time in my life. Juggling college, writing, publishing, and teaching, I was working 100 hours a week. Mentally and emotionally exhausted, I was left beyond any use. My body and spirit were ragged. Rest had become more than a desire, it was a necessity.

I had accepted the invitation. I saw the gathering as an opportunity to unplug, rest, and connect with like-minded people. Upon my arrival, however, I soon found myself alienated by several people at the camp. Simply put: It was the old schoolyard quandary; I just didn't fit in. I will admit, part of my isolation during the trip was of my own doing. As a youth, I was utterly open to sharing my private ponderings but that part of my character has suffered its wounds; leaving me a guarded adult. Being as private as I am, I do not make friends easily. I regard opening myself to another person to be a sacred act; it is not something I do on a whim. Trust must be earned.

While the trip wasn't the opportunity to connect with others that I had hoped it would be, I decided to use the time away to connect with myself and with the land. Instead of spending time in conversation with others, I spent time in the calm trying to hear my own thoughts, which had been drowned in the daily rush. I was gentle with myself over those days—I took long hikes; explored a new landscape that offered everything and asked nothing; sat by the fire late into the night; sipped mead under the stars and carried my journal everywhere I went—holding it as I would the hand of a sympathetic friend.

Autumnal Evening

The acorns fell like heavy rain—
Future giants falling to earth as newborns,
One day to stretch between the worlds.

The chipmunks chirped like cardinals
Back and forth at one another
As they raced along the ledges
Of the channeled slate walls.

Bees hovered in mid-air
With Zen like peace.
While the crickets chanted their mantra
Unto the harvest moon.

The red kernels of the burning oak log smoked,
Blessing those who stood around,
Witnessing the cremation of its century-old life.

The trees shed their leaves
Blanketing the path ahead
Like flower maids spreading golden petals
Before the bride as she walks to her union—
Before me, as I walk deeper into this wood
And offer myself as a companion
To the spirits in the surround.

A Progression

*Pennsylvania, Pocono Mountains,
Autumnal Equinox 2010*

If what we believe in is capable of change—growth—then our beliefs too must be allowed to evolve. There are ideas that do not change. (The philosophy major in me can explain.) You see, in philosophy, these immortal concepts are called "Platonic forms." They are perfect embodiments that remain the same no matter what shifts around them. For example, the idea (the form) of perfect *justice* will never change; it is eternal. As is the idea (the form) of *goodness* or *innocence* and so forth.

Concepts—perfect thoughts—do not change. They are, what they are. Yet if what we place our belief in is alive and capable of change we too—our beliefs concerning it—must also be allowed to change. Many world religions suffer from stagnancy; they do not allow for the evolution of knowledge. In the end, it is as though they regard God as a perfect *thought* that must remain fixed—frozen in time. The followers of these fixed doctrines are left with a belief in an *idea* of God rather than comprehension of the living being.

When we come to know one another, we must allow each other to change and become more, yet we do not extend this same philosophy to humanity's relationship with the divine. God (by whatever name you may know him/her) is an individual capable of growth; as such we who believe must accept that our knowledge concerning who and what the divine is must remain ever-evolving.

Truth is never fixed; understanding is *a progression*.

A Progression

I have made many journeys
Across many lands,
Over many lives.

In one life I was a Jew,
In another a Catholic.

In one life I was a Buddhist
And in another I was an atheist.

In one I was born a Druid
And in another I walked the shaman's path.

In one life I was a daughter
And in another I was a mother.

In one life I was a child
And in another I was an old woman.

In one life I was a warrior
And in another I was a pacifist.

In one life I was an idealist
And in another I was a cynic.

I one life I was a doctor
And in another I was a patient.

In one life I was a socialite
And in another I was a hermit.

In one life I was a reader
And in another I was a writer.

In one life I was a philosopher
And in another I was a fool.

In one life I was a virgin
And in another I was a lover.

In one life I was a servant
And in another I was a master.

In one life I was loving
And in another I was evil.

In one life I was frivolous
And in another I was practical.

In one life I was proud
And in another I was self-loathing.

In one life I was willful
And in another I was swayable.

In one life I had a family
And in another I was an orphan.

Throughout all these lives
I have been in need of acceptance.
Yet in the present life
I am confident in
What I know myself to be.

Always looking outside myself for knowledge,
I now look within.

In all these lives I sought
To connect to the divine.

Now I am at one with it—
Married unto it,
Like husband and wife.

Staying true to it
By remaining faithful to myself.

To gain knowledge of the divine
One must leave behind all religion.

To gain knowledge of the divine
One must part with the concept of God.

To gain knowledge of the divine
One must send away all intermediaries.

To gain knowledge of the divine
All we need do is come to know
Our original self.

Summer's Passing
Pennsylvania, Pocono Mountains
Autumnal Equinox 2010

At the end we reflect on the beginning. The autumnal trip had begun with a midnight train ride from the border of Connecticut and Rhode Island to Union Station in Washington D.C. and finally a connection on to Manassas, Virginia to meet the friend with whom I would attend the Pocono gathering.

I had deliberately chosen to take the late train to Washington. It is my belief that a lone journey through the night, without the distractions of business or technology, presents an opportunity to decompress and re-center ourselves.

In the dark of night it is easier to dream, to reflect—it is easier to be our true self. For years I have done what we are all forced to do—ignore my own wellbeing for the sake of making the monthly bills, to the detriment of my body, mind, and soul. At this point in my life, more than any other, I was in need of a pause—a time of nothing—during which I could put down my duties and let my poor mind rest.

Departing the train station at Westerly, Rhode Island at 11p.m. I churned down the coastline, through New Haven, through Penn Station, through 30th Street Station and ever-southward until at last, near daybreak, the conductor's call came for Union Station.

It was dawn when I arrived in Washington. My connection to Manassas would not arrive until 11a.m., giving me the time to check my bag and set out into the capital.

It wasn't hard to find my bearings in the unfamiliar city. I didn't need a map or G.P.S., only commonsense. The dome of the Capitol Building lay on the horizon; all I need do is walk toward it.

I will unabashedly admit that I proceeded into the heart of D.C. staring in wonder at the monuments, looking at it all through a Capra-esque lens. Since I was a teenager I loved historical American epics and Aaron Sorkin's series *The West Wing*, which I suppose romantically colored my view of the city. All the same, I was quickly relieved of my naïveté by the M16-armed guards walking the perimeter of the State Building and the abrupt pace of the city at morning rush hour.

Urban manners and automatic weaponry aside, I enjoyed walking The Mall up to the Washington Monument. I blocked out the din and simply tried to be present in the moment. I wouldn't have time to visit the museums or walk all the monuments; The Mall and back was all I would have time for that morning, so I opened myself to absorb as much of the surround as I could.

Ironically, it was the morning of the National Book Festival. The white tents were being raised on The Mall. Anyone who knows me, knows it is hard for me to pass up the smallest of library book fairs so passing up something of this magnitude was just painful. For a fleeting moment I thought about staying before reality sunk in and I pressed onward.

Morning joggers passed me; in the backdrop I heard the rush of the cars, horns, and shouts. I passed the Capitol reflecting pool and slowly progressed down The Mall. The National Gallery of Art, the National Air and Space Museum, the National Museum of Natural History and the Smithsonian Castle, all the while, the Washington Monument grew larger on the horizon.

It was a humid morning of haze and fog. The flags gathered round the monument were still in the breathless wind. I walked up the hill to the monument, came to the top and looked across—over the city. I stared at the Lincoln Memorial, tempted to walk on but, turning to my watch, I saw my trip was coming to an end. There was no time. This was my one moment to take it all in. I stood there walking the circle, burning the panorama into my mind then started back.

My morning sojourn was finished. I returned to Union Station to find my departure listed on the ticker-board. I collected my bag, paid the clerk who had kept it her over-priced fee, and boarded a train heading west.

A few short stops later and the train pulled into Manassas where I was greeted by the warm face of my good friend. Having stayed up the entire night, I was quite useless the first day in town. Conversation, dinner, and early to bed was the fate of what remained of the day.

Following a night's sleep, we loaded the car the next morning and set out just before dawn, aimed for the mountains of northeastern Pennsylvania.

As the sun started to rise we entered the Blue Ridge Mountains, and I learned firsthand that they had not been given their name arbitrarily.

The veil of morning fog hid the detail of the tall horizon. Only a faint blue shadow of the looming hills could be seen rolling across the landscape.

The road wound its way northward, cresting on the peaks of the mountains, then descending. The silence in the car during this part of the journey reflected what both my heart and my dear friend's knew: In the calm of the early morning our path was more than a pleasant ride; it was a spiritual venture into the sacredness of nature.

How ironic that the journey to the spiritual retreat turned out to be more enlightening than the gathering itself. Then again, I am not the first to realize that *the journey is the thing.*

Summer's Passing

Sparks from the fire
Were cast into the sky—
    For a moment able to live
    As red stars of the Milky Way.

The balance tips
And we pass into the darkness.
The days of long-daylight, spent.

The trees surrender their leaves,
Laying themselves bare—
    Exposed nakedly
    Unto those who dwell around.

The clinging leaves
Ripped away violently
From their mother bough—
    Orphans falling to earth;

Some to wither where they fall,
Others to have their mulchy ashes
Spread across the four directions—
    Borne away by the gusts of wind
    Unto a new shore.

Across the Distance
Connecticut, October 2010

At times, those who have the truth we are in need of are not physically nearby, so the lessons must take place from *afar*. We are able to learn from those who are at a distance by way of *the connection* we each have.

This connection flows from soul to soul—running through the fabric of the unseen. Every living thing is part of this *circle*. I am connected to you, and you to me. Through the connection all things can be shared—love, insight, healing, memory—spoken in an inaudible language that is heard with the receptors of the soul.

We are connected to the divine, to each other—to those passed and those present. We each have a connection to what came before us. We are each connected to our ancestors who, though they have left this world, live on *elsewhere*. And finally we are each connected to the divine. We have the means to explore the greater power from *within*—through reaching out into the unseen from our heart, directly unto the oldest consciousness. Where we can then listen to the truth firsthand.

The way of life we have at present does not encourage us to develop our connection to the unseen. The pace of our lives in this age is not conducive to maintaining a higher consciousness. Humanity suffers from the two extremes: Exhaustion and boredom. We fear life lacks any deeper meaning and so we fill it up with shallow busywork to distract us from the fraying holes within us and create a restless existence.

We fear the silence, as we equate it with emptiness. Yet it is only when in the quiet that we can hear the whispers coming from within. It is only when living at a moderate pace that we can maintain the proper level of consciousness and ordered priorities, so to be aware of the things of deeper significance occurring around and within us.

Across the Distance

I write this to you—
    You who read this long after
    I have departed.

I feel you tonight.
I feel you moving towards me.
I feel your eyes passing over my words.

The ink is still wet
Upon the leaves of my journal
Yet for you these words
Have long been set down
And begun to gather dust.

My future is your past
Yet there is a time
Beyond me and even you

And it is there dear kindred one
That our paths shall converge
And we can share a life together.

These words have connected us
Across the distance that time creates.
This night, we have met,
Though we have never touched.
And from this moment
You shall be with me and I with you.

We heal from our despair
Through learning

We feel nothing alone.

We are healed of our fear
In knowing that
We face nothing alone.

Know this,
You who are distant
Yet close to my heart:

If your search for
Understanding has brought you
To open this book
And read these words sympathetically,
Then know that along your path
You have found a companion in me.

Know that wherever I am.
When your eyes trace the outline
Of these words I can feel you
And I shall look up from my work
And send my love out to you
From across the distance.

Luminaria
Sturbridge, Massuchusetts
Publick House Est. 1770, December 2010

Early December is by far my favorite time to travel through the Northeast. The Colonial character of New England shines brightest in the early winter.

Winter in New England is a mosaic of luminous pieces. It is a warmth felt about the chest—a radiance. As gifted a wordsmith as I am, I cannot convey a New England winter here in this small space and do it justice. I find, all I can do is conjure the images and memories I have gathered over the thirty winters I have spent here and leave you to assemble the great picture in your mind.

When I think of winter in New England I think of an old fashioned holiday season. I think of the places I have visited: Walking through Mystic Seaport—the dirt paths gritty with crushed clam shells; moving through the rooms of Orchard House; walking down behind the Old Manse to stand on the banks of the Concord River on a cold morning; the inviting garland-strung door of the Concord's Colonial Inn; the nights of solace taken at the Publick House; the lantern-lit stroll through the grounds of Old Sturbridge Village, and finally the smell of evergreen in my family's home after we have set up the tree.

Reaching back into my memory smells, sights, tastes, and sounds flood into my mind. I hear music—bass, strings, and bells. I smell the hearths as I walk through downtown Mystic; I smell coffee and hot cider brewing on an open flame in a cast iron kettle and I taste a baked apple hot from the oven soaked in melted butter and molten brown sugar.

If ever there was a substance capable of healing the jaded heart it is that of the first snows to fall over the landscapes of New England.

One of my favorite places to visit just before the holidays is the Publick House—a historic inn located in Sturbridge, Massachusetts. Visiting this hidden gem of a place brings me back in time to 1771, the year in which the inn was opened. I imagine a time of home and family—a period of innocence, anticipation, possibility, and wholesome comfort for which we all inwardly yearn. Situated along the old Boston Post Road, the Publick House has been a solace for travelers such as George Washington and Benjamin Franklin. Visiting the inn has become a tradition of mine for three years running. Come autumn, I always make sure to book a room in the historic section of the inn for a weekend just before the holiday season.

Walking past the evergreen wreaths and yards of garland, the candles flickering, and the green, red and gold glass balls reflecting the light from a bright fire, I enter into the embrace of the deep oak-paneled surround. I walk across the wide floorboards to sit fire-side and enjoy a supper of roast duck and dressing in the old tavern.

While I am there for those few days life is as it should be. I give myself over to the period and let my imagination roam through the halls of the old house.

Anyone who has worked with me knows that I never rest while there is work to be done. Yet not many of my colleagues know *why* I work so hard: The desire to have a house and a small bit of land to call my own. I enjoy the Publick House as much as I do because for a brief time, as I walk through the halls of such old homes, I am able to spend a fleeting moment with the reality of the dreams I work so hard to realize.

During my days at the inn, my evenings are spent by lantern-light, seeping in the warmth of the gentle surround. Sleeping soundly, I wake in the dim hours of the morning and let myself drift for a time in the inlet between dreams and reality. For a fleeting moment I am not in an inn but in my own house. Among all the dreams I hold to in my heart, that of a house—a true period home—a few acres and independence is the one I simply cannot relinquish.

After a while of floating in the ethereal womb of this blissful thought, I sneak downstairs to the bakery and return to my bed with a cup of rich coffee marbled with cream and a piece of pastry in-hand, which incidentally is the only way to begin on a cold morning.

Indeed, some of the fondest memories of my life are centered around the early winters spent in New England. I do not observe Christmas or Hanukkah—my reverence for the season is not religious; rather, I celebrate the season of home and family when, after a long season of planting, learning, growing, and harvesting, the family returns to the home for a season of togetherness. As William Blake said, "In seed time learn, in harvest teach, in winter enjoy."

Luminaria

During the dark half of the year
The light that emanates from the home
Must illuminate our gray world.

The sun grows faint
So the hearth must glow bright.

The winter brings no warmth
So we must take refuge in the arms of loved ones.
The earth, in its respite, cannot nourish us
And so we must nourish each other.

Finished attending to summer fields
And autumn's harvest we return home
To the company of those for whom we work so hard.

Self-sustaining is the warmth and wholeness
Love brings to the hearts of those filled by it.

The Late Train Home
Boston, Massachusetts, January 2011

When I plan a journey, I never mind long travel times or layovers. Most cringe when I tell them I recently booked tickets for a seven hour journey with two layovers but to me the journey is dearer than the destination.

Some find their solace along the beach, drowsy under a warm sun; I find mine in a window-seat, traveling along the rails. No phone, no computer, no rush, and no concerns. I sit withdrawn from the chaotic world, quiet within myself, watching the landscape scroll by. Day moves into night and I descend deeper into myself. The horizon swallowed by night I see my own reflection in the glass now—a silhouette cast by the overhead lamp. The rhythm of the wheels churning provide an orderly comfort to a mind overwhelmed by a modern chaos.

The Late Train Home

The river of black glass
Cuts starkly through the
Blanket of cotton snow.

The wind is visible today.
The icy dust highlights
Its fine edges.

My time of meditation
Takes place on travel.

My mantra
Is the rhythm
Of the engine's movement.

The prayer wheel turns
As the train wheels churn.

My green eyes stare impassive
At the scrolling screen outside my window
As the mind's eye turns inward.

There is something holy
About traveling with a purpose
That serves something greater than self.

There are those who speak of religion
And those who carry a spirituality.

Do not speak
Of what you believe;

    live it.

During the Long Day Over the Sacred Night
Connecticut, February 2011

Like everyone, I have, at one time or another, lost who I am in the relationship I am in. From time to time, in my desire to be loved, I have altered who I am and conjured my own delusion. The most notable lapse was my first true relationship, he was 26 and I was 25 though as immature as we were we may as well have been 10 or 11, as lucid as we were we may as well have been drunk.

Hardly conscious, we never truly saw each other—not really. I saw him as what I needed him to be and he saw me as what he needed me to be—both of us trying to *mold* one another rather than *know* one another. There was no malicious desire to distort each other; rather, we each were desperate for company in a lonely time.

Eventually, the desperation lifted and we saw one another for the first time, fully conscious. The facades we imposed upon each other lifted when the despair that compelled us to construct them did and standing face to face without illusions, we found we were not a match. We surfaced from our despair; we sobered, the engagement was broken and the high hopes of two youths came crashing down.

In our loneliness and our need for love and approval, we sell ourselves for so little. The life of the modern man is an evolution of vices. From youth into old age we swap one dependency for another. Never knowing ourselves without the distortive presence of detrimental yearnings.

We need not content ourselves with being alone; rather, we must learn to bear being alone without letting loneliness drive us to reach out to anything or anyone in desperation. When we are desperate, we are willing to abandon ourselves if it means the end of our suffering.

When it comes to it, it is quite simple: You cannot be loved for who you are unless you are willing to actually *be who you are.*

During the Long Day, Over the Sacred Night

I lost myself
Somewhere between
The dawn and dusk.

Somewhere,
In serving you
I lost my self-worth.

Somewhere,
In trying to survive
I sold what gave the days meaning.

Somewhere,
In seeking the truth
I discovered the extent of lies.

Somewhere,
While needing your love
I altered myself to gain embrace.

Somewhere,
In needing to believe,
I let myself be deceived.

Somewhere,
While telling my story to you
Your fictions became my biology.

Yet somewhere,
Sometime...somehow

I found myself
There, in-between
Sunset and sunrise.

There,
In the dark,
When your light
Was no longer upon me,
I could see myself.

There,
In the silence
That opened while you slept,
I could hear myself.

There,
In the solitude,
Detoxified of that need for your touch,
I could feel myself.

There,
In the gulf between worlds
Where body and being are in harmony,
I remembered the whole of what I am.

There,
In the absence of desperation,
I gave myself the freedom to be
And came to love what lies within.

Departure From Avery Point

Long Island Sound, In-route to Flat Hammock Island, April 2011

D*eparture from Avery Point* was written from the deck of the Enviro-Lab vessel as it crossed Long Island Sound. I was accompanying the students from my sophomore Oceanology class to a bird sanctuary named Flat Hammock Island. Over the voyage to the island the students talked amongst themselves and I had a moment here and there to reflect and take in the sea air. That morning, I found my mind wandering to my childhood....

As a native daughter of Stonington and Mystic—old fishing villages in southeastern Connecticut—the thread of a seafarer-life has ever been woven through my life. My earliest memories of the sea are gathered around the Mystic Seaport Maritime Museum—a fixture in our small town.

While my mother was forced to keep a tight budget during my childhood, she always made sure that we had a membership to Mystic Seaport. On a difficult day, a walk across the grounds, a ride down the Mystic River on the steamboat Sabino or walking the decks of the old wooden whaling vessels was enough to make us happy.

The grounds of Mystic Seaport have soaked in the brightest days of my childhood. There isn't an inch of the place that isn't imprinted with some memory from my life.

When I was a child and even to this day, there are certain places that draw me when I pass through the gates. The Buckingham-Hall House is one of those places. It is a lovely old home built in the 1830's. To me, it is the epitome of all that a New England home should be: simple, well-crafted, and warmly inviting.

The old house certainly shows her age in places; the wide floorboards have grown parched and gray with time. But those very same years are what make her so endearing. She sags here and there as we all do late in life; the low ceilings made from the Colonial method of horsehair plaster make the space cozy. A steep narrow stairwell leads up to a small set of bedrooms that now overlook Chubb's Wharf where the Charles W. Morgan (the last wooden whaleship in the world and the oldest American commercial vessel still in existence) now makes berth.

The heart of Buckingham-Hall House or "the buck" as it is called by the staff is rightfully: The hearth. The kitchen is what gives the Buckingham-Hall House its character. It is not grand by any means; it is a small practical room. A wooden table sits in the middle of the room; resting atop it are handspun earthenware pots and pitchers. Nearby there is an old hutch that stores spices and dried herbs. The hearth itself is wide and always ablaze no matter the season. Fruit pies bake in the autumn and stews slowly simmer in the winter. The mouth of the hearth is haloed by a stone floor, which tappers back off into wood around the backdoor that leads out into a charming garden.

The smell of the hearth at the Buckingham-Hall House churning with the sea air will, for me, always be the smell of New England—of my home. The memories I have made at Mystic Seaport run like a thread through my childhood and into my adulthood: The Christmas trees tied to the mast-tops of the Charles W. Morgan and the square rigged vessel Joseph Conrad; a bowl of warm soup ladled from an iron pot cooking away on the hearth at Spouters' Tavern; eating a lobster, corn on the cob, and a bowl of chowder while sitting on the docks; the sound of sea shanties carrying from the old bearded man singing *cockles and mussels, alive, alive-o*.

It is both odd and extraordinary to think that I grew up exploring the lower decks of whaling vessels built some two-hundred years ago. I do not think I could ever live in a town that didn't give me a view of tall masts. I think the fisherman of old would agree that full canvas, makes for a full heart.

Departure From Avery Point

Setting out across the deep landscape—
Afloat atop the fathomless.
The bow presses forward
Into the misted Long Island Sound.

The Herring Gulls scavenging the dock
Are spurred into the air by the ruckus
As the sluggish boat rounds the jetty.

Barn Swallows curl through the air—
Plummeting as if to dive;
Only to pull up at the last moment—
Skimming the surface of a world
Into which they may not cross.

A flock of inky Cormorants
Perched on the stubby rocks
Are drying their wings
In the morning breeze.
    Arms open they face east
    Reciting their morning prayers
    Unto the listening wind.

The shadow of the strand
Fades in our wake,
As we proceed
Deeper into the mist.

Engulfed in the fog
It seems as though our boat

Makes no progress—
    Same black waters,
    Same opaque surroundings.

In a limbo of thought and motion
I stare transfixed at the striding crests.
A foghorn moans across the gulf.
Hardly reaching my ears—
    My mind wrapped in the depressing curtain
    Of the seemingly unchanging scene.

Until,
Jilted from the shapeless in-between,
Our boat is beached upon
The sands of the other shore.

Leaping from the bow
There is a crack underfoot upon landing—
    The shore is made of shells...
    Hollowed out remnants of feasts past.

Oyster Catchers hurrying along
    —peeping back and forth—
Discussing us featherless foreigners.

Black Backs sitting atop their kelp nests,
Each warming a clutch
Of speckled olive and brown eggs.

Smooth white breast puffed indignantly,
Drip of blood smudged on the lower beak,
    Red eye rims glaring,
The fearless mother sits,

Harking and barking,
Guarding the forthcoming.

Walking the rim of the island
We find ourselves back at the beginning.
The day wanes too fast.
The fog burns off in the mid-day sun.
The captain motions to us dune-walkers.
It is time to return.

The isle I am on
Merges with the isle on the chart.

The duties on the mainland call.

And I, unable to go deaf.

The Lament of the Wayfarer

Connecticut, July 2011

Often, when I ponder on grief, I think of the Greek epics and the voyages that separated lovers—one stranded on land and one set adrift upon the sea—both longing for one moment together yet unable to have it. Like Penelope in *The Odyssey* waiting twenty-five years for her husband's return, holding a belief that she shall see her Odysseus again long after everyone had abandoned the notion of his return.

All grief is suffered upon an epic scale. When loss strikes, our world ends. There is no greater struggle than that of recovering from an apocalypse of death.

Having suffered my own losses and walked the long procession, I know there are no words to express the depth of the heart's mourning. It is a woebegone eternal night of oblivion unto madness. Our life is smashed against the rocks of a bitter sea and we are left gasping for breath between strangled sobs.

The Lament of the Wayfarer is a reckoning. It is a promise between the lover and the lost that there shall be a reunion and, once had, there shall be no next parting.

The Lament of the Wayfarer

When the day comes
And I at last clear this dense wood,
I shall meet you on the other side.

When the day comes
And my path comes back to the place it began,
We shall go on to that next place together.

When the dawn comes to this night
And I have seen you through the worst,
You will sit with me until I close my eyes
And I wake in my bed,
in that home I left so long ago.

When the day comes
And I can at last pull up
The moorings holding my soul in place,
I shall journey to you
And set fire to this vessel I dwell in

Never to leave you again.

The Lowest Point
Boston, Massachusetts, July 2011

I write spiritual verses rich with ecological references. This paints an image in people's minds as to what I must be like. After reading my work, most tend to envision me as the equivalent of Laura Ingalls Wilder sweeping over flowered hilltops in a long sundress. When I reveal to them that I grew up a raging tom-boy in a suburban slum, they are a bit taken aback.

I found my love of nature while exploring the small bit of wilderness near our neighborhood. It consisted of a few acres of wood behind the local high school and a small drip of a pond. I spent most of my early childhood here—knee-deep in muck, stalking frogs and painted turtles, catching catfish and building forts.

Our neighborhood certainly wasn't a *ghetto* when compared to those harsh areas in the inner-city. It was more the equivalent of *the other side of the tracks* where the poor kids lived. Drugs, gangs, even murders had been carried out in our community. Yet despite being raised in such a harsh place, when the time came to venture into the city for business matters as an adult, I was utterly out of my element.

I am not a city person. As much as I have experienced a *street upbringing*, I am a small town girl who is bombarded by the city.

When reflecting on what leaves me so unsettled when I venture into the city, these are the conclusions to which I come: The city, I suppose, is meant to be a pinnacle in mankind's achievements; however, in my view, there is nothing less civilized than an urban area. Its character, to me, has always been one of rudeness. The sounds, the sights, and the smells are offensive to me. In my mind, *civilization* equals peace, enlightenment—a transcended way of living. There is nothing peaceful or harmonious about a city.

I write this on a train, returning to my small Connecticut village after a day of business spent in Boston.

The Lowest Point

Drunkards passed out,
Mouths hanging open
As they lie sprawled out
On the stained benches.

Waking long enough
To kill off the plastic bottle
Of cheap booze,
Drool on themselves,
Then passing out once more.

Passing cars kick up
The murky water in the streets,
That smell of sewage
In the humidity.

Cities—mankind's accomplishment—
Are barbaric outposts;

While true civilization
Is found in the places man
Has yet to touch.

Humanity need not create civilization;
All we need do is not screw it up.

The Truce
*Arcadia State Park,
Rhode Island, September 2011*

Dreams are the gestation of a future reality. We do not come into being fully formed; rather, we gather, build, and grow. So too our matured identity—what we will be and do in this life—grows as well. Our reality begins as aspiration—vague dreams that sharpen over time until at last tangible. In nourishing our dreams we enable our future self to be born. In protecting that which is still taking root, we allow beauty to enter and flourish in this world.

These were the thoughts gathering on the afternoon we [Andy and I] walked among the tall pines at Arcadia National Park near Richmond, Rhode Island. It was our first trip to the preserve. After a day of rambling in search of the trailhead, the long hot car ride had slightly drained our adventurous spirit. We found ourselves longing for a shaded spot with a cool breeze; what we found was much better.

It was around midday. The road arced and the car climbed a steep hill. It was then, as we came to the crest and began the journey down, that we found Browning Mill Pond stretched out below us. Grateful for such an inviting rest stop, we parked the car in a small dirt lot near the pond and were finally able to stretch our stiff legs.

We walked down to the water's edge. Looking at it one wouldn't think to call it a *pond*. A pond to me is mucky, small, and is home to frogs, turtles and catfish. Browning Mill Pond was a basin of deepest blue dropped into the center of a dense pine forest. No, this pond was a lake at heart. As wide as many of the lakes I had seen during my summers spent in the Finger Lakes region of Upstate New York.

Directly beneath the sign welcoming us to the pond was the unwanted site of a "No Swimming" sign. But it was in vain. The sweat from the hot day had made us blind. The sign said *no* but the cool inviting water whispered otherwise. We convinced ourselves that no water this blue was meant to go un-indulged. Looking down shore, I saw that we weren't the only ones inclined to ignore the signs; there were many natives gathered on a thin strip of shore bobbing and diving.

For hours Andy and I swam. Several times I waded out far from the banks—standing in the shoulder-high waters—staring out across the rippling plains, unto the houses nestled along the distant shores, dreaming of what could be.

The Truce

Pluck a strand of wind
And listen to the trees quiver.

Run until your heart pounds
And watch the stagnant surface
Of the pond ripple.

Throw back
The suffocating blankets of false comfort
And let yourself feel the renewal of the rain.

Only when we overcome
Our fear of being alone,
Can we come to know the company
That is always with us.

In surrendering, we are cradled.
In accepting, we are able to impart.
In kneeling, we stand taller.

Gather what is worthy of your devotion
And never betray it.

So that, in the end,
You will know that,
Though you be small,
You poured out all that you are
Into what is greater

And in doing so,
Became a part of it.

Withered

Connecticut, During Hurricane Irene, August 2011

In late August 2011 New England braced for the landfall of Hurricane Irene. We battened down the hatches, the storm hit and Connecticut suffered power outages for several weeks.

This made for an interesting week. At home we made French press coffee by sterno and enjoyed games of chess and Monopoly by lantern-light.

When I was a child, it used to take at least a category 3 for us to lose power, but now it seems we lose our lights at the smallest gust. Now, even a good nor'easter leaves us without power for a few days.

While to some it may be a mystery as to why we lose electricity during relatively minor storms, I have walked through the woods of my home state and know part of the answer.

Many of the trees throughout my hometown and throughout Connecticut for that matter are brittle—they are standing but they are dead, which presents a problem for the power lines, seeing as so many of the roads throughout New England are still wooded.

A few years ago I started to notice the change. It was after a trip through the woods I used to play in as a child. I had gone back to walk amongst the trees I had grown up with only to see that, while they were still standing, a great number of them were indeed rotting.

During those weeks we were without power in the wake of Irene, I watched the light and power companies blame the long outages on the numerous downed limbs and trees and I started to feel grieved for these natural places that were soon to be gone. Not just the woods I was raised in, which were now dead, but the greater scale of ecological crisis taking place globally.

The storm passed, power return, and New England moved into autumn. I stared out at the trees lining my backyard, waiting for the foliage to turn, only it didn't. The wind was so strong it had bruised the leaves and the trees withered without bringing their autumn colors to the landscape. That year summer withered without a graceful transition from green to golden. It seemed a great shame. During those weeks, the grief for all that is happening to destroy the irreplaceable landscape seemed to come into sharp relief.

Withered

Autumn without color
Is death without transformation.

The storms that raged over the summer,
Wrung out all color from leaves,

Leaving them bruised upon the bow
And our world bleak and bereaved.

Blue Mornings on the Concord River

The Old Manse, Concord, Massachusetts
November 2011

In late autumn of 2011 I had the great fortune to be invited on a trip to Concord, Massachusetts, arguably the epicenter of contemplative literature in New England. From the moment the trip was planned I felt as though I was making a literary pilgrimage. Concord was the birthplace of the Transcendentalist movement, home to Emerson, Alcott, Hawthorne, Fuller, and Thoreau. I felt rightfully overwhelmed.

When arriving in Concord I was struck by the history of the place. It was the little things at first: The old graves, the slate sidewalks, and the Colonial farms. Then, there was the first site of Orchard House, which called to my fond childhood memories of enjoying *Little Women*. The road signs pointing the way to Walden Pond, the Old Manse, Emerson's home, and Minute Man National Historical Park, where the opening battle of the Revolutionary War was fought in April of 1775.

Not at all surprising to those who know me, my first stop upon arriving in town was the public library. As I walked down the stone path I knew I was following in the footsteps of those New England minds I most respect. Opening the narrow double doors, I proceeded through the short mud room and into the main room where I was suddenly struck by the history of thought held within that small space.

At times when I am hiking through the woods behind my Connecticut home, I will ruminate on the history of the ground beneath my feet. Wondering if perhaps some *Mashantucket* or *Eastern Pequot* village might not have resided nearby and if the paths I walk weren't once crossed by one among the tribe.

I found myself asking the same such questions that morning

while exploring Concord. As I walked the paths I felt Henry and Louisa brush past me in the crowd. As I grasped the old wrought doorknobs, I shook hands with the past. Using the window of their books, I had ever-been looking in on the lives of these kindred minds, but finally, on that day, I found myself invited in from my musings.

Walking across the slanted floor of her room to sit at her small desk, I found Louisa. Roaming the hidden paths behind Orchard House, I found Nathan. Reading in the study of the Old Manse, I found Ralph Waldo. Sitting along the Concord River I found Margaret. And there, along the banks of Walden, I found Henry.

Blue Mornings on the Concord River

Across the yard of the Old Manse
    Following the stone wall
    Down to the boathouse,
I come to the edge of the river;

Its gray, still waters
Mirror the marbled sky above.

The bare trees
Are stiff in the cold breeze.

Thick, stout bushes are scratched
Into the scene—
    Etched there by the pallet knife
    Of the *great painter.*

Plump geese waddle along the rim,
Sifting seeds and bugs
From the muddy grass roots
Passing through their black beaks.

Just beyond
    —Across the arc
    Of the wooden bridge—
There grows the meadow
Of the minutemen.

Running through the draping grasses
The farm boys fired their muskets.

    Son against son,
       As blue collided with red,
The world changed.

In the fields
Along the Concord River,
Violence begot a nation.

A century later
The passion continued to
Pulse through the place
As the minds in the Manse
Spurred a revolution of intellect.

How small am I to stand here
Along the banks of a river
That has seen so much.

I who but scribble in the margins
Of those classics penned
On this ground.

In the distance
The sun is rising
Above the treeline.

The field is smoking
As the morning fog rises.

The rifles sound…
The hearth of the Old Manse smokes…
And the river rushes ever-on.

On the Far Side of Walden
Walden Pond, Concord, Massachusetts,
November 2011

Visiting Walden Pond was a surreal experience. The trip granted me one of those rare opportunities to step into the reality of a place I had only ever imagined. I hold the same reverence for the grounds that most reserve for holy sites in the East. One can sense that something happened on these grounds of significance. It wasn't a great battlefield or the resting place of a prophet. Instead, the events that unfolded here were intellectual. One man sat along the banks of the water exploring himself and the sacredness of the world he was born into.

There is a narrow path that follows the brim of the pond. At several points along the trail you will find stairs created by flat rocks leading down into the water. They lay like a spiral staircase of flat rocks set into the steep banks. Each time I came upon one of these waypoints I paused. Something about these stairs was haunting to me. I felt as though, if I followed them—if the decent was made—perhaps some renewal might await in the blue depths. The land's offering of a natural baptism to the traveler—a renewal that needs no overseer or ceremony to give it weight.

Walden is important because it represents the simple man's journey unto profound realization. Feeling the need to simplify and meditate, Thoreau made deliberate choices to clear for himself a moment in time—those seasons he spent along the banks of Walden.

We all have our Walden; we each have a place willing to cradle us—to protect our quietude, push back the trivialities, and give us a chance to be fully conscious. In this age of streaming information there is no quietness left. The constant din of the media babble and endless menial tasks fill our precious days. No, in this age, we must *make* quietness; for it does not occur naturally in this world we have constructed.

We each have the ability to follow the example set by Thoreau and countless other unnamed souls throughout history, to make the choice to clear away everything else for a time and explore this world and our place within it. Personally, I find such sojourns necessary. I will not come to the end of my life having never stopped to truly examine the world I am in and the beliefs I hold.

Move off the path—descend—and see what renewal lay in the depths.

On the Far Side of Walden

Ankle-deep in mud
Along the banks of Walden,
I find my footing.

Standing in the ruins of the cabin,
I return home to meet the brother
Born before my time.

We souls close in ideals
But distant in years
Keep council together.

When one of us passes along
The next will pick up the thread
And carry on the thought.

Walking the rim of Walden
The wheel of my life takes a turn.

The waters are a mirror
And the banks a respite.

On the far side of Walden
One can look out across the wide waters
And see the world reflecting.

It is a place in the journey
Where one can take solace,
Pause and look back with clear perspective.

Coming to the end of the path
I am not who I was
At the beginning.

Self-sufficient
Connecticut, December 2011

A stable house cannot be built upon the shifting sand. The mind needs stability—a set of fundamental truths that it need not question. Without this foundation the mind has no anchor to reality and you find yourself adrift in the unknown, bordering the surreal—lost in the gray between what is real and illusion—unsure what is genuine.

Nothing can be built without this foundation—we cannot begin to trust any truth come to without first securing this baseline knowledge.

Years ago, everything that I held to be true from my religion, to my personal relationships, to my preconceived notions of reality, was shaken. The upheaval, while difficult, brought about needed change; however, as long-time certainties were pulled out from under me, my mind did struggle.

It is courageous and wise to demand that the truths we are given be justified, no matter the proclaimed authority of the sources. Doubt uncovers lies and in this skepticism can indeed be a virtue. Yet when you call into doubt the entirety of your world be sure to redefine the foundational truths your mind rests upon or else sanity may be crushed in the collapse of your world.

Self-sufficient

I drown in the what ifs and whys,
Live off hopes
And reach for belief.

Years spent waiting
For a God,
For a lover,

For a father,
For a mother,

For a savior,
For an answer.

Thinking myself incomplete
Without any or all.

The time has come to explore
All that is within this one soul.

...All that I am,

Rather than musing further
Over all that you could make me.

The Boat Rocker
Connecticut, December 2011

As an adolescent, I was fearless. I followed my heart without regard for consequence. Looking back, there are times when I considered this aspect of my character a strength and times when I considered it a weakness.

Often, when looking back on my younger self, I see a fool ignorant of the workings of the world, and yet there are other times when I look back and see myself as wise in what truly mattered; for while I was ignorant of the backlash that might follow certain choices, I knew enough to heed my heart above convention or expectation.

I was once a boat-rocker; I was never one to *sweep things under the rug* or silently accept an illusion or false version of events, no matter how much those around me wished me to be otherwise. This angered a great many people around me who depended upon the illusions for their daily comfort.

Slowly, I have come to terms with the fact that eventually, we boat-rockers must face the wrath of those who reap from the status quo. Few of us survive with our defiance intact. At some point, all those who fight a long battle look back on the clarity, strength, and bright-eyed hope they had in the beginning and long for renewal.

There was a time in my life when I had fierce clarity for more than a fleeting moment and during that time I acted without fear; however, circumstance has since taken its toll.

I am exhausted now, inside and out, after being overworked and made to struggle to have the simplest of things. I have been sedated by my uncertainty and fatigue, but oh how I long to return to that fearless time when, no matter the storm churning around me—no matter what forces it meant I would face—I rocked the boat if it meant a lie would be overturned.

The Boat Rocker

I overturned
The boat once
And nearly drowned.

Coughing up
The heavy waters
I choked upon,
I lay limp.

Rescued,
Part of me died
In the storm.

Gasping, the defiant self
   Was the victim
And the pragmatist
   The survivor.

A boat-rocker once,
I am left to fear
The open waters,

Stranded on
The barren island
To which my trauma
Confines me.

I stand quietly screaming,
Yearning for cause
To revive the wild ways,
To take a stand as I once did—
   Fearless and fierce amidst churning swells.

Fleeting Moments of Fierce Clarity
Moving Mandala of the Soul
Connecticut, January 2012

My spirituality is a braid of many faiths. Raised a Catholic, I spent many years studying the Catholic/Christian traditions including the apocryphal gospels. Eventually, I moved beyond this doctrine and went on to study Judaism, Buddhism, and shamanism before finally moving into my own spirituality.

Hints of each faith I studied have stayed with me over the years. One need not be a part of a religion to respect its tradition. One such set of teachings I admire is Buddhism. I do not consider myself a Buddhist; nevertheless, I believe there to be wisdom in the tradition. This poem sprang from one such Buddhist practice I respect: The making of a *mandala*.

The first question asked by those who read this poem is: What is a mandala? The sand mandala is a Tibetan Buddhist tradition involving the creation and destruction of murals composed of colored sand. Monks and nuns work on a mandala for several days—devoting great effort to its design and creation. Then, when the mandala is complete it is ritualistically destroyed—swept away. The sand is then gathered and poured out into a river or ocean, where it will be distributed by the tides and currents—released throughout the world, acting as a blessing. The creation, destruction, and scattering of a mandala symbolizes the Buddhist doctrinal belief in the impermanent nature of all things.

Fleeting Moments of Fierce Clarity—this work's namesake—is my realization that the process of defining ourselves is much like making a sand mandala. We define the edges of our soul—the aspects of our identity—only to evolve. While it might seem we are periodically destroyed by hardship and change, we can, in time, discover it was not a violent end; rather, it was a death that led to a needed rebirth.

Fleeting Moments of Fierce Clarity
Moving Mandala of the Soul

Years spent defining this self
   —Carving out the edges of this mind—
The channels of this spirit.

Now to take that sculpture of identity,
Which over time becomes the mask,
And shatter it.

To achieve perfect knowledge
Of the soul within—to experience
That fleeting moment
Of fierce clarity.

Then to surrender.

To grasp the known
And then transform again
Into the unknown.

The grains of sand
From the mandala of soul
That take one form

Are scattered—swept away

Then regathered
To make a new form.

Definition and evolution.
Empowerment and surrender.

We come into knowing
Then set back out as the student.

The child and the elder
Are one and the same.

We are free
Only when we allow ourselves
To be boundless.

Do not drive yourself insane
Longing for the destination.
Live while on the journey.

Deep Notes, Quick Pace
Connecticut, January 2012

Deep Notes, Quick Pace is one of those poems I am tempted to let the reader interpret without a relating-point. The emotions that triggered the creation of this poem are so full they cannot be wholly captured in words.

The poem began to gather one evening while I was alone listening to a piece of modern orchestral music. Alone, working long-past night and into early morning, a particularly moving piece of music came over my headphones. I ceased tinkering with the document I was working on and opened a note pad—feeling the words taking form. The music was so radiant—so intense—I found myself transcended; I was taken away from this flat earth and into a place of suspended bliss. I became acutely aware of the miracle of drawing a single breath. Riding a wave of poetic ecstasy, I savored the moment and as my appreciation for the experience ascended within me, I found the joy slowly muting into grief.

This poem embodies *a fleeting moment of fierce clarity* as it captures one of those brief, transcendent moments of pure love and appreciation that, when experienced, affirm to us that the struggles in our daily life are worth persevering through; for there is a beauty in this life to be experience, the sweet glimpses of which outweigh the heaviest sorrows.

Deep Notes, Quick Pace is an expression of the ache we feel when the music heard is so beautiful yet also bound to fade into silence as it inevitably approaches its end. ...Those delicate moments of radiance, when the moment is so precious and the soul next to us so dear that we hold tight—willing the moment to last—knowing that it will fade. These moments are so beautiful that we mourn them, even as we live them.

Deep Notes, Quick Pace

The despairing beauty of the music
Causes us to lament life as we live it.

The capture and loss of the moment
   —The bittersweet quality
      Of these fleeting days—
Deepen the significance of what is felt.

The fragility of time
   —Strong one moment,
      Fading the next—
Is the seduction
Of the human experience.

The momentary ripeness of this flesh,
The short space of this breath,
Quickly withered,
The beauty is missed
If not lived.

Each note
Sweet and somber
Has its harmony.

This movement
Rushes so quickly
Unto silence.

Fall backward
Into the tide.
Be carried away.

All or None
Connecticut, January 2012

I have a very close relationship with my mother. Years of joys shared and decades of difficulties faced have made us tight-knit. Our bond has evolved over the years growing from mother and daughter, to sisters and best friends. In many senses, her and I only have each other; having no other family of which to speak. So this too—facing the world as we two alone—has bound our small family quite strongly.

My mother has faced a great deal of adversity in her life. Since I was a small child I have watched as she confronted a gauntlet of hardship. She is a deep yet simple woman, with simple dreams—many of which have sadly yet to come true.

For many years now, she and I have faced the gauntlet together. We struggled together, cried together, hoped together, and reached for our dreams *together*.

In recent years, however, there came a time when it looked as though our paths might diverge—where we might not endure all things together.

As I experienced some modest success as writer and publisher, I found my life changing. My landscapes of possibilities widened. With my success came new choices—choices to go off and follow a few of my long-held dreams; however, only I was given this offer, not my mother.

The reality that my mother and I dreamt of knowing—had held and protected together—was offered to me and me alone. During this time I found myself faced with a choice. Would I go and live the dreams she and I had held without her or let the moment pass—holding out for an opportunity wherein we both might experience that reality together. Fully knowing that a second chance might never come.

My mother, of course, did not begrudge me the chance to live my dreams—it was not about suppression or spite but rather, it came down to this: Is achieving our dreams a matter of simply experiencing that dream ourselves or reaching that sought after reality *together*? I asked myself: Was it enough to have the dream or didn't some part of the dream's full fruition lie in the fact that both my mother and I would experience that happy life together?

I made my choice. I decided it was *all or none*—either both of us would emerge from the difficulty or none of us would.

An illusion of disconnection is causing a great deal of suffering in the world. This mistaken idea that humanity is composed of individuals existing separate unto themselves rather than small pieces coming together to compose the body of humankind.

The *all or none* mentality is what we need at present. Drawing a line and stating: None shall eat until all can eat; none shall have shelter until all have shelter; none shall drink until all have clean water; none shall take comfort until the sick have been attended to—that the world will stop until the state of matters are improved for all.

Some might call this view naïve but this sentiment—this loyalty to one another—is what will save us. In the end, none shall truly succeed unless all succeed. There is no going on if a member of our family is left behind, and indeed we are all family.

All or None

I feel survivor guilt

As I stand successful
Passing through the ward room
Filled with those undone by
The poison of decayed dreams.

We live with our dreams
In life and in death.
Bound to our best parts,
When they die, we die.

The random selection
Of dreams allowed to live
Unravels those
Whose longing is neglected.

In my success I find isolation;

It would be better, I think,
To be swept down the river
With the tide of the forsaken

Than be the lone one
The inexplicable fates
Condescend to spare.

Resented by the disregarded dreamers
My achievement is bittered.

We go forward as one
Or not at all.

The Voice
Connecticut, February 2012

So often in this life, when we are in need, we look outside ourselves for answers. In search of *truth* we explore new faiths, we go on pilgrimages, we read books, we listen to secondhand accounts of insights, and so on—always reaching into the world outside ourselves for truth that may be relevant to our life alone. As though we believe our soul to be disconnected from the greater meaning underlining this world, forcing us to wander in search of it.

So many of us feel orphaned by the divine. We spend our lives in mourning, feeling that we have no place of belonging. We begin this life afire with belief, conviction, and certainty, only to lose such feelings as we proceed deeper into our adulthood. World-wearied, we come to the conclusion that we have lost our faith. Blaming the difficult circumstances of this world for jading us, as though believing that time kills faith.

But we are wrong. It is not that our belief has died. The true problem is, our belief has been *drowned out*. It is not that we have *lost* ourselves, it is that we have been *consumed* by a way of living that distorts the shape of our soul and dulls the workings of our mind.

We need not seek the answers concerning our place in the greater workings outside of ourselves. We are not unknowing; we are not lost. Our problem lies in the fact that the inner-source of answers we each hold has been pushed down into us by all the trivialities present in the way of life humanity has chosen.

We don't need to fill ourselves with the insights and philosophies of others in order to learn what is *beyond the veil*. We don't need to go out; we need to go *in*. We don't need to fill ourselves; we need to let what is within us rise to the surface.

The fast-paced way of life is not conducive to consciousness. We become more aware of what is truly at work in our lives by pushing back all that hinders us, all that influences us, and all that exhausts us. We move from youth into adulthood—we merge into the current of this fast-pace world and our mind is consumed. Our days are taken from us as we stress over a thousand little things. And then, at the end of the day, we are so exhausted by the *trivial matters*, we don't have the mental power to contemplate the *greater matters*.

We connect with our answers—our truth—by pushing back all the trivial things and create the quietness needed to listen to what already lies within. The most spiritual act one who is searching for insight can make is to do nothing—to clear a space of time in which to listen, and to see, and to think. No work, no social engagements, no all-consuming social media, no media streams—nothing but yourself and the quiet. The goal is to arrange your life in such a way that you are protected from all that would drown out the higher workings of your mind—to leave behind all the entanglements that rob you of your consciousness each day.

Our most intimate moments—our clearest periods—are spent when we are alone within our thoughts be it while taking a long hike, soaking in a warm bath or curling up in a cozy chair with a mug of hot tea.

Some of us fear silence. In the quiet our mind whispers hard truths and we retreat into the roar of the chaos to spare ourselves from having to listen any further.

After running from task to task it is hard to sit still. When we withdraw from the current, it can feel like a detox period as we gradually leave the fast-paced way of life. We are left not knowing what to do with ourselves.

The only way we stop running is when we are forced to—when we are sick, when we lose our internet connection or when we leave our cell phones at home by mistake. But quietness can be a deliberate choice—a chosen way of living.

The transition from the fast to the calm can be difficult. Our world contracts but in time we adjust to not being a part of the chaotic current. We grow to appreciate the slower pace. We come to value what we already have instead of yearning for what we don't need. And it is there—when all that would divert us or exhaust us or deafen us is gone—that we find the connection we have within us running right to the source of the wisdom we seek. It is there—in the quiet—that we can finally see and feel and think and hear and understand.

We wander, looking for the face of God. It constantly eludes us, not because a *greater power* is absent, but because we are moving too fast to see it. We endure each day in a silent crisis—not knowing who we are—because we are too exhausted and wearied to even be conscious.

The Voice

Come not unto me softly.

The pace of my life quickens;
The wind, so fast now,
As I sprint through my days,
Numbs my skin.

You cannot whisper,
As you did,
In the slow, warm
Days of my youth.

The din of the wind
Raging in my wake, deafens.

The fast scroll of the scenes
Along the road, mesmerizes.

The nagging of the trivial
Leaves no time for the vital.

You must be the force
That brings me to a halt

Loud enough to carry over chaos.

You cannot hesitate.
You cannot be shy.

Catch my eye,
Tug my sleeve.

I implore you—
Break upon me.

Rise within me like a flood tide—
Washing the clutter in my mind
Out to sea.

Blow me off this course
And back out into
The unmapped places.

Renew the curiosity
That once propelled me
Beyond the confines
Of the drudgery.

Authenticity
Connecticut, March 2012

The constant presence of influences rob us of our opportunity for authenticity. From the moment we leave the womb we are bombard with influences. As children we are shaped by our parents, our friends, our teachers, the media.... These dominant presences sway our way of thinking and perceiving. Then, later in life, our addictions and detrimental needs tend to dictate our identity and actions.

For all the talk of the pressures society puts upon us, I find it easier to ignore popular opinion when compared to finding clarity and freedom from my own ill-emotions—my own detrimental yearnings. I am greatly influenced by my emotions, both for the good and for the bad. My passions, curiosity, and ideals compel the actions and thoughts that define me as an individual. Yet I am also influenced by my fear, my fatigue, my doubt, my insecurities, my loneliness, my regrets, and my desperation. These neuroses grip all of us at times, altering and influencing us in ways that lead us further away from our true self.

To perceive, to speak, to hear, to act, to think—all without influences to sway or alter us. Who might we be?

To live life without any and all influences and in doing so experience a moment of authenticity wherein we might know who we are.

Authenticity

I shall take years
Fill them with solitude
Draw from that basin of quietude
And with the clarity wash away
The ill influence from my life.

Before I pass from this place
And am reborn into a fresh unknown.
I want to achieve authenticity.

I want
To know who I am.

To know,
Not who you made me
Or what circumstance has left me.

To shape my own words,
No longer regurgitate yours.

I want
To hear my own voice;
For I have talked

But never spoken.

The Mystic River

*Along the Mystic River,
Mystic, Connecticut. March 2012*

An early morning meeting cut short brings about a day without walls. Venturing down to the water's edge, settling dock-side with a coffee and a new notebook, no creature was ever more content than I.

The first day of warmth and light since winter. I sit on the edge of the dock like the turtle sits on her rock, bathing in the sun.

Never before did I feel as though I had a place of belonging but when I found New England—when I discovered her character—I felt the peaceful embrace of a long-awaited welcome.

My love affair with New England is like that of a couple who were forced into a marriage who grew complacent with each other, only to wake twenty years into their life together to realize how they had grown to love each other.

As a teen I raged, restless within the confines of a small town. In my twenties my focus turned to the landscape of this soul. But now, on the eve of my 30th birthday, having returned from the inner-pilgrimage and rebellion of youth long-since resolved, I see the face of my motherland and find comfort and pride in being a daughter of New England.

The Mystic River

The slow current of the river
Is a balm for the overworked mind.

Watching the continuous flow
Of the ripples,

My eyes sliding out of focus
As the din of the day is drowned.

The scene seeps all stresses from me
Carrying them away

Down towards the bridge.

Simplicity
Stonington, Connecticut
June 2012

There are days when life is complex and I require some answer to the hows and whys that surround the mysterious forces at work in my life. And still there are days when life is simple and all I need is a heavy coat to wrap around me, a sturdy pair of boots, and a bag big enough to carry my load.

There are those days when, emptied by the hardships to befall us, we require a great deal for happiness. Those dark days when the worth of being alive eludes us in our haze of weariness. But then there are those days when small joys bring a warmth to our numb heart. Those days when our suffering recedes, the gray curtain rolls back, and the beauty of what it is to draw breath pervades. Those days of gratitude, appreciation, and *simplicity*.

Simplicity

There are some days
When it is good enough
To simply walk out my door
And be warmed by the sun.

To watch autumn bleed into winter
And feel the warmth return
To remind that part of me
That always forgets that
Spring will come again.

CONCLUSION

This Fleeting Moment of Fierce Clarity That is Life

In the end, the small experiences—so sharp at the time we live them—blend away, leaving us with an overall impression of the moment rather than the vivid memory of the hours spent. ...such is life.

Perhaps, when our last breath is drawn, we will see that this life has itself been a fleeting moment of fierce clarity.

Life: A period of years that passes in the space of a breath, during which we become all things.

Birth unto death is a fiery ride from the light of first dawn through the blackest night. The heart of history beats and in that space of a moment we are born, live, die, and come to dust. Through this human condition, we are sent naked into a world of violent experience. For one fierce fleeting moment—during the flash of this single life—we are all things: We are passion; we are need; we are desire; we are gratitude; we are despair; we are hope; we are transcendence; we are rage and reckless abandon; we are hunger and we are fullness.

Through this life—the duration of this body—we take voice and express the pent-up needs and agonies of our mute soul. Alive we burn with fervor—self-consuming—mortals demanding answers of the god.

Transient spirits passing briefly through this body, for a short time we can reach out to another—bind ourselves to another; lose ourselves and find ourselves; cower in shame and stand in pride; feel the grip of deepest love and burn till the new dawn extinguishes the night.

Some are undone by the moment, unable to bear the neck-breaking swings from happiness into desolation. This fleeting moment of beauty is so profound, we grieve when we reflect on it. This moment of joy, need, curiosity, and loneliness is so entrancing that we forget we are experiencing it all in a free-fall.

Even the most desperate pain—the most terrifying madness—has a sweetness; for in feeling, we know what it is to be human.

Enduring the tempest, naked amid the surging elements—this is life. To be vulnerable and afraid; lost and longing for home, lying in need of some other from who we may draw strength. In the torrent of experience we are swept up into the currents of emotion, left grasping at each other for some anchor, else we be carried off.

As with any fleeting moment, the sweetness is in the sorrow and the sorrow in the sweetness. The moment comes, it fills us, we are alight with it, and then...it is gone.

A CLOSING NOTE

I nearly named this book *Fleeting Moments of Fierce Clarity: Journal of a New England "Contemplative"* harkening to the Transcendentalist circles of previous generations. In the end, *Fleeting Moments* is not as much a picture of New England as it is of myself—a contemplative poet living in New England. I am a composite of my beliefs and my surroundings.

As I prepare to offer my reflections to the world, I do so with a hope that my sentiments and truths will be received by open minds.

This book, in the end, became its own fleeting moment of fierce clarity. As you read it you glimpsed a flash of my own soul. I hope you learn something of yourself from all that I have offered; for, sharing such intimate reflections is no small thing for any person.

Society pulls at the hem of the artist's robes—begging that they be made to feel something in this numb age. And if we [the artists] can overpower our insecurities and make it out onto the stage we just may give you the pause you seek. Artists and poets are like streakers—strange introverted streakers—who, overcome by the muse, strip themselves naked and dash through the world... through a moment of your life.

The last page turns and here—on the Connecticut coast—my day goes on. Rooted in the old salt marsh, a lone cattail stands, skin broken but still intact, waiting for a stiff wind to take it into its next life; lingering, broken, bent but not yet free.

Cool vapors of the nearby sea float in the air. The winds ride the crests in to shore. Water and brine—salty capped waves—clear the sinuses and soul all at once. One stiff breeze and the seeds are cast unto the tides of the world.

Acknowledgments

This work is dedicated with gratitude and affection to:

Andrew, your love and support has both widened and filled my world. From Arcadia, to Walden, to Southbridge, and to those places not marked in miles, we have traveled a great distance together.

To Ian Marshall for agreeing to write the foreword to these wanderings. And to: Philip F. Gura, Chris Highland, Keith M. Cowley, and Frank L. Owen for offering advanced reviews.

To Michael C. for your friendship and for making the autumnal journey through the Blue Ridge and Poconos possible.

To The Thoreau Society, the staff of Orchard House, and The Trustees of Reservations for maintaining the paths the great poets cut.

To the dedicated staffs of the Jacob Edwards Library and the Westerly Library for their encouragement early on in my career.

To the Publick House and Concord's Colonial Inn for their lodgings during our cold winter journeys.

To Mystic Seaport for the sanctuary.

To my dear friends and fellows: Frank, LaRue, Theodore, Todd, James, Jason, Mathew, Anna, and Jeremy.

To those kindred hearts who go unnamed for the acts of kindness and love. (You know who you are.) Your presence ever-reminds me that a single act of compassion can change a life.

And finally, with special gratitude to my mother for her unwavering love, support, friendship, and guidance. We make the journey together.

About the Author
L.M. Browning

L.M. Browning grew up in a small fishing village in Connecticut. A longtime student of religion, nature, and philosophy these themes permeate her work. In 2010 she wrote a Pushcart Prize nominated contemplative poetry series. Following in 2011 with the release of her first full-length novel: *The Nameless Man*. She is a graduate from the University of London and a Fellow with the League of Conservationist Writers. Balancing her love of writing with her passion for publishing, she is a partner at Hiraeth Press, Associate Editor of *Written River*, and Founder of *The Wayfarer: A Journal of Contemplative Literature*. In 2011 Browning opened Homebound Publications—an independent publisher of contemplative literature based in New England. For more information visit: www.lmbrowning.com

Foreword by
Ian Marshall

Ian Marshall is a professor of English and Environmental Studies at Penn State Altoona and a former president of the Association for the Study of Literature and Environment. He is the author of *Story Line: Exploring the Literature of the Appalachian Trail*, *Peak Experiences: Walking Meditations on Literature, Nature, and Need*, *Walden by Haiku*, and *Border Crossings: Walking the Haiku Path on the International Appalachian Trail*.

HOMEBOUND
PUBLICATIONS
Independent Publisher of Contemplative Titles

GOING BACK TO GO FORWARD is the philosophy of Homebound. We recognize the importance of going home to gather from the stores of old wisdom to help nourish our lives in this modern era. We choose to lend voice to those individuals who endeavor to translate the old truths into new context. Our titles introduce insights concerning mankind's present internal, social and ecological dilemmas.

It is the intention of those at Homebound to revive contemplative storytelling. We publish introspective full-length novels, parables, essay collections, epic verse, short story collections, journals and travel writing. In our fiction titles our intention is to introduce a new mythology that will directly aid mankind in the trials we face at present.

The stories humanity lives by give both context and perspective to our lives. Some older stories, while well-known to the generations, no longer resonate with the heart of the modern man nor do they address the present situation we face individually and as a global village. Homebound chooses titles that balance a reverence for the old wisdom; while at the same time presenting new perspectives by which to live.

WWW.HOMEBOUNDPUBLICATIONS.COM